For Lyn

Best Wishes

Penelope Shuttle

April 88

Penelope Shuttle (signature)

PENELOPE SHUTTLE

The Lion from Rio

Oxford   New York

OXFORD UNIVERSITY PRESS

1986

Oxford University Press, Walton Street, Oxford OX2 6DP

Oxford New York Toronto
Delhi Bombay Calcutta Madras Karachi
Kuala Lumpur Singapore Hong Kong Tokyo
Nairobi Dar es Salaam Cape Town
Melbourne Auckland
and associated companies in
Beirut Berlin Ibadan Nicosia

Oxford is a trade mark of Oxford University Press

British Library Cataloguing in Publication Data
Shuttle, Penelope
The lion from Rio.
I. Title
821'.914   PR6069.H8
ISBN 0–19–281974–7

Library of Congress Cataloging in Publication Data
Shuttle, Penelope, 1947–
The lion from Rio.
(Oxford paperbacks)
I. Title
PR6069.H8L5 1986   821'.914   86–2562
ISBN 0–19–281974–7 (pbk.)

Set by Promenade Graphics Limited
Printed in Great Britain by
J. W. Arrowsmith Ltd., Bristol

Contents

The Hell-Bender

It is an hour without heroes in an Ohio valley.

The hell-bender is there,
'a large aquatic salamander
about eighteen inches long and very tenacious of life'.

He is a summer beast
nimbly folding the water into shapes
that suit him,
his garments he might sleep or hunt in.
All feebler things are his serfs, his fodder.

(Your sigh glitters like sun on rapid glassy water,
in the valley the campers deliberate over maps;
the morning trembles in its hurt.)

The salamander pours himself through his waterfall prairies;
garden pools do not admit him
but anyway he, supple stone of blood, fiery rope,
streaks away from their logic,
away from the white proud superstitious flowers,
away from the hooded lips of the snapdragon.

He does not recognize his children
but shoots tumultuously through the water
as if his sister were with him.
He is the serpent of sun who lives in water,
he is a master of water and fire, cool heaven and hot hell;

He is clarified to his length, his spare eloquence.
He knows he will never die.
(Who said he is only a kind of fish?)
He darts through water like thread through a fine needle eye,
he is very tenacious of life.
He can bend hell.

Snake

Snake lazing in the wet grass,
less useful than cow or horse.

Line of silver on the family path,
silver as the Rio de la Plata.

Serpent silver as my ring, my bracelet,
laughing silently as those two circles.

Serpent tingling from place to place,
one of those who do not save lives,

at whom the countrywomen fling sharp stones,
but whose daughters greet with sudden smiles.

Creature more magic than mouse or rat,
more thoughtful than donkey or cat,

whose cry is mistaken for wind in the trees,
from whom so much has been stripped,

now you are only one limb,
one skein, one thumb,

you are a long thin silver skin,
a rod that works for god.

Because of your perfection
we say you possess venom and deceit

but whoever has perfection
can do without compassion.

Silver female with your nest of pure white eggs,
you live both by basking and gliding,

you die without screaming,
you come to an end,

your silver stiffening to pewter,
then thawing back into the shallows of earth.

Your young wriggle free,
bruised but seamless,
each one her own stepping stone.

The Vision of the Blessed Gabriele

(*Carlo Crivelli, National Gallery, London*)

In the evening sky, swallows

and the saint in his robes of evening cloth
gazing upward with his worried stare.

Is it because there is no star?

His feet have slipped out of their sturdy medieval scholl sandals.
He kneels on hard sand where thin grasses fountain
and starfishy cacti flourish near a few egg-shaped pebbles.

The frail tree that for years has borne no plums
touches both the saint's shoulder and the sky.
He is holding his hands palm to palm,
making the old holy arch of fingers and thumbs,
his two little fingers making an exact oval.

He looks up at her as if she's a trespasser,
hanging there in her larger oval in the sky,
the queen and her babe,
as if he sees her as the queen of untruthfulness.

How worried and angrily he stares at her,
his hands kept holy and invulnerable,
his bare feet ugly and ordinary, a man's feet
on a man's earth,
behind him the barren tree
and above him, she and her fertility.

Swags of fat fruit, unbelievable ripeness, loll across the sky
supported by an old ragged linen hammock;
hanging from the sky not stars but outsize heavenly fruit
knotted in a casual arrangement of dirty bandages.

4

On the sandy ground his holy book lies open,
forgotten, its script of red and black abandoned
as in horror he stares up at the fruit,
apples and pears from a giant's orchard.

Who put them there, apple and pear,
growing on the same branch, fruit bigger than a child's head?

The hedgehoggy halo of the saint quivers.

Within this glistening vagina the sky has blurted open
like an eye or a fruit, there is this queen or golden doll
carrying her stiff golden child,
golden and ruby-red couple in the sky,
cargo lugged along by cherubs, the crumpled robe
of the woman evidence of their haste.
They peer round the edges of the mandorla, singing a suitable song.

And like a gulliver the helmeted man
with his thoughtful grieving head
lies face down on the path in the wood,
alive or dead, who knows?

Not the saint, still staring up at the sky with its storm of fruit,
at the mother of gold, her foot set on fruit,
on another goddess's golden apple.
The child holds either a second golden apple or maybe a golden ball.

The saint gapes. This is the pain of the answered prayer.

In the pond by Gabriele's feet, in the green water,
the drake moves to the lustrous duck
with almost unnoticed longing, with vigilant love.

On the branch of the plum tree, a bird is about to fly away,
north to China or south to India.
When the bird has flown the saint will be able to weep.

The Weather House

(*for Peter*)

I usually understand you
when you are working with electricity
because we have often run away together
into the park of storms
where the thunder and his sister lightning live;
there the clouds come to us like pets,
eligible grey mammoths asking to be fed and groomed.
We build our weather house
from the shaking white boughs of electricity;
the branching sky alive with the sleepless storm
is our garden where we gather flowers of fire and hail.
When we fear our life is slipping
back into familiarity and calm ground
we return to the special house with its trembling galvanic rooms,
to the garden seared with the tallness of trees,
to ardent air prickly with hope of rain.
How the clouds crush us under huge pigeon-grey feet
before releasing their naked furnaces of rain on us,
till we are like fountains kissing!
How the storm aches with its own fame, its long steps
pouncing to reach us!
Electricity wires us, it shoots its fix into our veins
and our dreams lengthen into flooding weather, the sweet breath
of downpour, the waterfall gasp of it.
I usually understand you
when you are working with electricity
and despite the shocks
I clasp you in my arms, our skins jolting with the power,
sharing the voltage,
Storm the friend and lover of our hearts.

Clouds

The long hair of the world
rises in defiance of gravity,

the world's waist-length hair
flying, levitating,

skimming the tree-tops,
the world's runaway hair

lifting its chastity
high into the air,

the world's hair
halfway between heaven and earth.

And the wind, blue experimenter,
sleeping upon it.

Selena

('her cherry's in sherry'—woman's period)

The unripe cherry has the luxurious bitterness
of the earth's satellite, the scarlet morals of it,
its acid blotches stinging your tongue;
a moon cherry that mulishly leaves its flavour
in your mouth all month; and that secret early woman
in the sky, whose soft authority will not fly away
but who holds us in her strong birth-marked arms,
or hides us behind her natural naive skirts; that taste
of sharp cherries steeping our tongues only means
we are her namesakes, the selenas . . .

She is the poverty of an unimportant person,
a boy, say, or a simpleton, one grown but not
and never to be adult;
she may be persistent as a child that sucketh long,
or as reproachful as headmistress; her green-goose
ceilings and her books of grief are all her;
on these observatory nights when the taste of oboes
blots out what you said or might have said to me,
she falls in scalding rain, she approaches the waning sea,
accepting without protest her unprotected position.

She stands before us, a sudden window, an intense door;
though she lets nothing ever be quite closed or quite open.
There is always one more letter to be learned
in her alphabet, one next fruit to be tasted;
after the sour text of the cherry, the golden and tiring orange,
the juicy pang of the scarce pear, the pearly apple's
pedestrian-calm;
the clingstone peach with pink and velvety skin;
her hands offer a midnight feast, her bride's charm.
She is almost capsizing with the fruit she brings.

8

All her roamings lead her to richness,
a richness that, as soon as it's at its height,
begins to diminish, little by little, until it becomes
another currency, another night's work in a sky so intimate
it reaches the most sensitive part of the world, a leaf maybe,
or a fish sleepless in his ocean,
or a pillow-slip blowing in the wind that has waited an eternity on
    the line
for this touch of moonlight on the worn white tucks and embossings
    of its linen,
selena's touch, her concubine's breath,
her fruits and their felicity.

Disdainful Jack

(after the painting, Our Jack, *by Henry Scott Tuke, depicting Falmouth-born Jack Rolling on the quay punt* Lily, *off Custom House Quay, 1886)*

The boy in the bright blue coat,
navy-blue cap jammed on cropped head,
right hand hooked loosely in the rigging,
boy with his sad monastic look,
his uncomplaining expectant stare,
his knowledge of blue.

Behind him, over his shoulder,
rises the terraced harbour town,
its royalist church and roughcast houses.
Its unsheltering streets
only numb or vex him.
He has no time for the land, nor it for him.
Ashore, he trips facetious in clogs,
but on deck his feet are light and undisabled.

With little humour and less hope,
he stares disdainfully at the painter, at us.
He dwells on the voyages he's chosen,
the hammock he'll be suspended in,
the little whips of salt already burning his lungs.

He has no time for luck.
His look tells us that, plainly.
He watches the waves,
his eye forming its own past tense,
looking beyond Tuke
far out to sea,
to the day or night of his drowning.

In blue jacket and blue hat,
dressed for sea and sky,
Jack's at a standstill, lounging stiffly,
prisoner of his own dignity.

He has no sweetheart.
He keeps his energy for knowing the sea.
His gaze is the narrow ledge along which he inches his way,
lonely but used to it, to the narrowness,
the fear of missing his step.

How the ocean will welcome Jack,
who knows the cues of the drama,
who appreciates the cabaret of storm,
the syncopation of the tides.

Only when the waves close over his head
will he smile, relax, at last open his heart,
find his expected home, his unexpected happiness.

Wise sailor, he never learned to swim
and so can sink without a struggle,
the fraternal waves letting him down gently
on the rope of his last breaths,
Jack safe forever now
in one of the galleries of the sea.

Horse of the Month

Here is a horse made of sleeplessness.
He is devoted to me.
I am sewn to his saddle,
am his established rider.

Breathing upon the sky,
the horse makes me love him.
He repeats his breath of flame.
The sky is burning, old shawl.

The trash and dust of smoke
is luck on our tongues.
The horse begins to speak, composedly.
We ride down green lanes, clover byways.

I ride him like jewels.
We wheel around our red-coral valley,
inseparable, sleepless,
grass turning to fire wherever our hooves of blood fall.

Outdoor Anniversary with Maria

(wife, husband, and their child, in the garden on the tenth anniversary of their marriage)

WIFE

Indoors, I am official custodian of our museum.
Marriage only ripens outdoors, or in the greenhouse
among lazy plants with real and undisguisable names,
fuchsia, idle orchid, where light sinks against the glass roof,
a warm breath approved by the sky lifting up its own lenient blue
    towers,
holding them high as the sweetheart moths;
while the tall historical stiff-bladed rushes
fringing the shallow garden pool do not move,
not even to rummage out one secret from their reflections
adorning the rusty lid of the water.
I am a wife, but I can punish terrestrial boys.
This garden is our planet; here, as husband, wife
and child, aided by interested animals, cat and hedgehog,
we go on counting out the autumns, and this is our tenth.

Our child is named after the open sea,
she comes towards us glowing with her emergencies,
her inch-of-rain demands. Maria.

HUSBAND

Maria! The en-famille of the years has taught us to shout
so that the silence of earlier journeys would seem shameless now
in this bronze-green garden of bracken and water-pepper,
of flowers with their aching seedheads and northerly perfumes,
bushes odd and sad as acquiescent machines in dreams
where the wide-awake airman flutters in his tangled strings.
Yet this garden has also been a ship to me, for sailing far
from the house, the exhaustion of chairs.
I know I listen callously to music indoors, Maria's
indolent childish practice that is nothing but egg-whites

13

and sugar beaten into sleep . . . I've been more at home myself
in the garden than indoors, all the confusion of pianos and pillows,
that dreamless cage of ours, the beds we enroll in;
only in the garden can I hear our hearts
beating behind the strong walls of our chests,
sense the blood circulating, not too fast, not too slow,
setting the pace at which we go.

WIFE

But you go round the garden like a nomad,
as if even here you had no home, your heart tensing
as if you could give up breath forever! Yet I know
you are untravelled, are just a stay-at-home, and your laughter
is what I like best in you still, you whose name
is folded up so tightly in me after all these years
that now it has taken the place of my heart,
it is your name, Michael, that beats and sends my blood
from head to toe, not the usual gruff thudding cardiac muscle;
deep in me, right now, your name is beating like a breath
from a sky so casual, so familiar, so ours that I think
it would be easy to fly up into, if I did not know so well
that we are earthbound, gardenbound, housebound, marriage-
     bound,
ten years spun round us in a cocoon, a fleece, a fleece
you saw first and wanted, despite the cost,
you did not care, just as now you don't care and dart away,
waving mockingly at me from the other side of the pond,
you with your insolent happiness, your grin,
as if the walls of our house were not waiting for us . . .

HUSBAND

Sometimes the house is a ship too, but a sunken ship,
a ripped hull lounging sideways, submissive, proud of its errors,
and we are the crew swimming forever
through baffling fathoms of speckled fish, giant undomesticated rays,
sleeping octopi, idle fluent sharks, you and I and Maria swimming
through the cabins and caverns of the wrecked ship,

and under the water all we say goes unheard,
it is the earliest of dramas, when words are not used,
and only gestures tell the tale of the storm,
the wreck and the days of the drowned as we learn
from the total loss of our strength all that is worth knowing
about the rest of our lives.

HUSBAND AND WIFE

In our garden-swamp of sage and fox-sedge,
we meet as searchers and collectors,
celebrating our marriage, anecdotic as dogs.
The child races from green slope to crimson hedge,
her bell-shaped blue skirt fringing bare knees,
girl with her seven shrieking giggling ghosts,
each one smaller than its sister, one for each of her years,
as she leaps, scything the rainy air with her arms
and cupping the long sheer sky in her hands.

WIFE

Above Maria's head, the day-flying moth of marriage
beats its wanderers' wings,
as happy in this mock-nuptial rain and unruly garden
as it will be sad inside the house
when the craft of cushions and the art of tables
shut all three of us in again, prisoners of a wedding.

HUSBAND

I enter the house almost boastfully,
as if I were your sister not your husband.
Poor house, it is so frail, so unheroic,
the rooms as tired as we are.
This is a house that knows how to welcome us.
It does not want to know what is happening outside,
its old floors are ready to be walked on,
its air wants to be breathed. Come in, Maria,
come in, Mary, let's pretend you are painting a portrait of it.
The garden is not a husband, or a wife. Come in.

MARIA

Like an owner I go from room to room,
playing tricks to keep off the sun
who tries to follow us indoors; I shut him out.
Today I know that I am what happened to love.
In the shuttered house I eat plums, soft sweet fruit,
my tongue and lips crimsoning with the juice,
daydreaming of the garden where my childhood ends.

Ashputtel

(*Cinderella*)

My face is black with ashes,
I can see nothing
though I hear the near-silence
and nostalgia of my father's fountains,
smell far off the festive bonfires.
It is pain without salves,
my body is metal,
slapped across my shoulders
I'll ring like a gong.
If there are treasures in this dark
I cannot hope to carry them off,
I know I'm uglier than my sisters,
will turn any beholder to stone.
Must I sit here forever,
tethered, homeless,
or dare I ask for a diet of light,
a shining nourishment,
rinse off my blindness,
see above me in their prosperous palaces
my sisters, my million sisters, the stars;
my Vanessas and Sophias,
suffer them to pour their nymphean pitchers
over me, clothing me
head to foot in their clean threshed glitters,
guipure of starlight at my hem,
crochet of moonstone and pearl at my throat,
my bodice and skirt whiter than the lace aprons of my stepsisters,
my gloves woven from the first memory of silk,
at my breast an unbarterable diamond;

feel again my stockinged-feet slipping into sandals that shine?

Fougou*

They left all their grief
down here in the shadow-luck
of the tunnel,
left the hunger of their lives,
left history to ripen without them,
and climbed on boneless ladders
to a place as real or fictitious as daytime dreaming.
Are they safe now?
Does death taste like the cream
off the top of the milk?
Or are they casualties of heaven,
restless in its light?
The rough ascent
out of this chamber
took them somewhere high up
and not known to us;
now they know the answer,
they are the wise ones now
even though they did not read or write,
did not linger
in any deepness of thought,
were cruel as children and as modest.

They are not castaways.
They hold the long juicy stalks of eternity in their hands.
They feast.

* prehistoric cul-de-sac chamber in Europe, used either for storage or ritual purposes.

Orion

Orion standing at ease
just above the horizon.

Offer the dreamer
a window
through which she may see
the most secret parts
of the warrior.

Orion,
the within-er,
the penetrator.

The god's cool semen
falls upon her,
he throws it upon her
as if flinging
the last drops of wine from his cup.

Her dream will come true.
A child is waiting for its life.

The thunder-lord of stars is wailing,
changeling caught by flesh again.

Girlhood

In the sky, the airship drifts
at the fingertips of the clouds,
unexpected and immodest.

The girl is pulling an earth
out of the sky,

her own earth,
her wild vehicle, her light-as-air thief.

The earth is her own
but by evening she has lost it again,
her womanly Zeppelin.

Her torch of breath lights up the house,
she loves her budding polished tables,

her chairs real as kings,
doors waiting for her love.

She calls her house,
'Waiting'.

In the sky, the evening airship gleams,
fuller, more difficult, less buoyant than girlhood.

Giving Birth

Delivering this gift
requires blood,
a remote room,
the presence of overseers.

They tug a child
out of the ruins of your flesh.

Birth is not given.
It is what is taken from you;
not a gift you give
but a tax levied on you.

Not a gift but a bout
that ages both the contestants.

Birthshocks hold on tight, for years,
like hooked bristles of goosegrass,
cleavers clinging to your skirt and sleeves.

The raw mime of labour
is never healed,
in giving birth
the woman's innocence goes,
loss you can't brush away,
it stains all your new clothes.

No longer can you be half-woman, half-bird.
Now you are all woman,
you are all given away,
your child has the wings,
can resist the pull of the earth.

You watch her rush up,
clowning her way through the cloud.

And you applaud.

Chrysalis

Like all mothers
I gave birth to a beautiful child.
Like all mothers
I wiped myself out,
vanished from the scene
to be replaced by a calm practical robot,
who took my face,
used my bones and blood
as the framework
over which to secure
her carapace of steel, silicon and plastic.
I was locked out of her clean carpentry
and smoothly-reprimanding metal.

Yet that robot's rude heart
flowed with love's essential fuel
because my child was one of the millions
of beautiful children
and knew how to tackle the machine.
She embraced the robot woman lovingly
each day
until her circuits and plastics wore away.
Now the soft real skin can grow,
the blood and breath move again,
the android is banished.

I emerge from the chrysalis
and go forward with my child
into the warm waters of the sea
in which we are both born at last,
laughing, undamaged,
bathing alive in this salty blue,
my motherhood born out of her,
her woman's name and noon out of me.

Child and Toy Bear

It is essential
to have the bear
in the bed
though he is nameless
and disregarded throughout the day.
At night he must lie beside her
so that she can sleep,
his black nose firmly clenched
in her hand,
the spar that keeps her afloat all night.

Miss Butterfly, Miss Moth

Butterfly and moth,
one primrose pale,
the second grey as god himself,
both dead,
the child keeps them
in a Flora carton
with air-holes pierced in it;
airy tomb,
plastic sepulchre
she has given to moth and butterfly
as a sanctuary
where they can find peace,
transform to their next stage
which, she sings hopefully,
(Miss Butterfly, Miss Moth),
will be fairy or elf
but fears will be only wing-tip dust,
a tick of mist;

for the child has undreamed her song before.

Bear-Hug

Childlessness crushed me,
a bear-hug

I never breathed
till I bore her

though now in her clasp
I hurt

being drawn so far
from my breathless life

Why compose
on a guitar
at six years old
a curious refrain
entitled
Horse Mane?

But she does

The Child

Elaborate manifestation of a smiling dog
clockworking his way over the carpet;
he crouches and barks in a high peeping metallic woof;
the child follows the toy into the room,
remarking calmly, he's really a hedgehog, you know . . .

2

The wind windowed me out
but I held on to my friend
or my friend held me with his teeth.

Did it hurt, I ask.
No, she says, it didn't.
Landing safe in the net of her dream.

3
Or she dreams of the cat,
his dovetail pleasures.

She dreams I was angry with her.

She dreams she had an egg
which hatched out into a nasty chick.

She dreams she was given another egg.
And all the other children had eggs too.

4
Another dream (she said)
was when I had a little cat
that got smaller and smaller
and ran into the bush;
when I called it
lots and lots of little dogs came out.

Sober dream dogs,
gentle canine companions.
In the child's dream they bark quietly.

5

As soon as she wakes
she starts up all her unmarried nonsense.
The rebellious parents
have to find all she needs,
her props;
they puff up and down stairs,
supplying dolls, pearls, frogs, rice, birds.

To the child nothing is a luxury,
every thing is a necessity.
For her, each day has the glamour of convalescence,
all conversations possess the repartee of scripture.

Masks

The child has masks.
It is easy to forget this.
Behind her masks
of today and tomorrow
is yesterday's face,
see, she is still too young
to understand anything
but food and sleep.
My threats are no way
to break her silences,
to curb her fires,
there must be a way
of speaking
that runs true and clear
from the womb's infant
to the child who faces the world,
her school masks of fear and pride
sprouting fresh each day;
she flinches but does not retreat;
she wears a bruised lazy-mask,
a stiff oldfashioned anger-mask,
one summer mask glitters, gifted with speech,
another is a poke-tongue laughter mask.
She has her heroic silver bedtime mask.
My own pedantic mother-mask watches.

There must be a language
for me to speak, for her to utter;
a language where the sweet and the bitter
meet; and our masks melt,
our faces peep out unhurt, quaint and partial as babies.

The Knife Knows How

(after tubal ligation)

The day after the knife
had written its scarlet line
just here down my belly,
making its little slit,
peering with its bright eye
into my womb,
playing lives and deaths,
the hot sky screamed with planes
displaying their skills
for the festive crowd,
warrior-wings criss-crossing
the coarse blue sky
with their knives of noise and vapour,
a sky sliced into segments
like a birthday cake for a child
who will not be born now
because the knife knows how
to open up a belly
and cut it free of a future
I cannot house or find a name for,
a life that would lie like lead
not flesh within me,
whose first cries would bleed
in me
like that giant kiss of crimson smoke
the planes carve in the perfect skyblue sky.

Act of Love

At night, riding our bed like a willing and dethroned horse,
we are secret depositors proud of our flaws,
flaws that scratch a diamond;
you are a stinging mirror to me, I another to you.
We are each a bird ruthless as cat
but we let that cruelty go into the dark
and lie lithe as lizards, side by side,
our fingernails extracting silver from our hearts,
the distinctive lode we work,
darkness arcing with our buck and doe brights
until we rest for a little, partners slumped on the ropes of night's ring.

Our outstretched arms anchor us, inseparable;
my nipple is hard as diamond, treble and desirous;
my breast-skin soft as unchaperoned moss;
your hand on it a serious shimmer;
my breast grows newer, newer,
my yanked-open laceless nightgown's bodice,
its cotton seam is caught tight around my ribs
where my heart is beating gravely and loudly,
its blood full of Burmese strength and mystery.

The night outside is a teetotal drum we flood into silence
as your delicate hard sex presses against my hip;
when we meet and join we hold our breath,
then breathe out all the burning novelty of our bodies,
a big vapour furling into the room, flag
made from our clear-sky flesh, our unearthly diplomacy,
our hauntingly-real fuck;
I watch my familiar but elaborately-lifted leg
misty and incredulous in the straight-faced dark.

And as we are not blind or dumb
this is the time we stare and cry out best
as we wear out our weariness with thrusting,

our eyes open and glossy, our throats humble with aahs,
sighing into inaudibility, our lips soft reams
of silence; we're giddy with our tongues' work,
as if two serpents had become brother and sister . . .

As we cast ourselves into the night and the act,
our smooth knuckles shine,
we are gasping as we smell the sleep to come,
waiting for us beyond this untraceable room;
now we clamber the summit of old-friend mountain,
rising faster in our clamour,
swinging locked-together in our bell-apple nakedness,
in the double-pink hammock of the night
made of touch and breath,
(the purposes of the engineering!),
a labour of love as we rush towards that trembling edge,
toppling over yelling into the fall, the rapids,
the waters we enter, fluid as them,
my sex hot and hidden, perfect and full,
the corner of our sinews turned,
a clear answer found, its affirmative leaping from our mouths,
my body's soft freight shaking and accepting
in the clairvoyance of orgasm,
and your answering sheer plunge from mountaintop
into river,
flowing where the bed was.

Sleep takes us then and drops us into its diocese,
drops us from night's peak into a dawn
of martial ardour, of trees mad
with old-clothes spite, a morning
where the starving still wait for us,
each with their lonely cloudy gaze.

So only the sugars of the night offer us any breakfast,
only our night's act of love feeds us,
the remembrance of our bodies like slow-moving turtles
lifted from the sweetness of a sea of honey,
flying into more sweetness.

Only the touchwood of our sweet bed
dams the savage sour torrent of the day.

In the morning we must say goodbye, not hello,
goodbye until the untouchable day has gone
and the night recalls us again to our study,
to our sweating gypsy-wagon sheets,
our navaho pillows and rich pastures,
the mintage of our wild skins.

Love the Children

Love the children
better than light
or the powers
of marriage.

The children
see through darkness
and the secrets of silence,
they are more than alive
even when they sleep.

Their perfect guesses
protect us from what we know best.
Their courage
floats on wings
like a summer
we must avoid
because it is too vivid for us.

We remain
on our narrow, cold fjords,
ice chaining our feet,
hands blindfolded, breath
white as the kindness of a flower.

Equestrian Order

Evening
for approving the absence of colour.

For the sadness of shoulders,
for learning a language
left behind when the ice melted.

An evening for sketching hedgehogs,
their learnèd bristles.

An evening when the wetpaint snake
slinks by
with his fast tenderness,
his lack of speech.

An evening I wish to give you,
but cannot.

Evening of balloons and dirigibles
floating slowly in the sky,
their big hollownesses
like departing souls,
sure of their destination;
sure of the air, anyway.

An evening for riding to the moon,
equestrian order;

I go riding on horseback
with the creatures of evening,
the shepherds and queens of energy,

riding with the carefulness of eternals,
beyond the soft fat of animals,
crossing bridgeless rivers into living midnight,

hearing my shadow cry out in astonishment,
'I can keep up with you on your flight!'

Different Trees

Willow and poplar, poplar and willow,
one bending low, the other erect,
both bitter, both bringers of fever,
willow and poplar.

The poplar's unlovable leaves,
the willow's too-female branches,
their surveillance of the afternoon,
roots sunk in earth's deep water;

if I am afraid now it is their doing,
an old fear
renewed under the supervision of these trees,
binding me, just as I thought
I'd made an end of customary terrors,
was growing strong and demure,
companioned by tamer trees;

big, safe, old and battered by boys' play,
a rough platform of planks
slung in the trunk's lap,
black tyres swinging on ropes from branches,
trees more like big old brown dogs than trees;

more used to boys than birds;
and honest where these others
are ill-omened and unkind to me
with their wind-whispered poverty,
the rough cat-caress of their tongues
mewing, 'hunger, hunger',

like the spite-spell of some obeying witch
who comes out of her black bazaar
to kill me with her cool amorous rumour,
her leopard's milky laugh.

The Martyrdom of St Polycarp

He had known John
and others who had known the Lord
but he was betrayed by a servant,

arrested late in the evening
at a farm outside Smyrna,
hens scattering in panic,
geese retreating angrily,
children peeping from corners
to find out who are heroes,
who are villains.

This happens around the year 155,
the arrest of an old man
who had known those who had known
the Lord,
had known John.

In the city
a crowd assembles for the games,
officials, wives, magnates, courtesans,
labourers, idlers, children, artisans;
animals baying, trumpeting,
the stench not a clean farmyard stench
but a festering stink,
the reek of a blood circus.

The old man and the proconsul converse,
they see eye to eye,
they are the only philosophers
within five hundred miles,
and able to bear their differences,
the roman reluctant
to condemn the venerable man
whose honour he can see.

The old man shrugs, smiles.

'How can I curse Christ,
for in all my eighty-six years
I have never known him do me wrong . . .'

And the crowd is yelling,
                   'Kill him,
he is the one who destroys our gods . . .'

Even the cripples and lepers join in.

The circus gods need blood or ashes.

So because he is commanded
the proconsul orders the burning of Polycarp

'and the flames made a sort of arch,
like a ship's sail
filled with the wind,
and they were like a wall round the martyr's body;
and he looked, not like burning flesh,
but like bread in the oven
or gold and silver being refined in the furnace.'

He was like bread in the oven!
Like gold or silver in the furnace!

He turned the torture circus
to a fiery circus of joy, flames of the spirit.

But the cruel spectators did not clap their hands,
or fall to their knees, or say to the children,
look, there is a miracle, a man alive in the flames.

Did the people say, have our gods done such things?

Did they warm themselves at those flames?

The old man stood
with the flames flowing round him
like a weir of fire,
sailing in his ship of fire,
safe in his tent of flames

as the outraged crowd damned him.

At a sign from the proconsul
(curious in private life
about the supernatural)
a bored boy-executioner
braves the miraculous ark of flame,
pierces the old man's heart,
freeing Polycarp,
who kicks his corpse aside
and becomes a soul
and the crowd go on cheering,
children laughing, the rubbish gods ungrieving.

Flaxen River

Why is the flaxen river with its sweet taste
tempting me again with its pomp, its fierce look,
its history and its air of a patron?

Why is the far shore a peach corner,
soft honey on the tip of a woman's finger,
the silent candour of her campaigner's smile?

Why is the name of the river lost,
like some palatial thimble or salt-spoon loggia,
way out of reach in these elixal waters?

Why is the river luring me to lie
in its cubicles and closets,
a cageling, a mogul child bobbing on its peephole currents?

Why does the river watch me
with its eye-witness longing, its riddle-wedding love?

Why is it tempting me to close my eyes,
to forget, to become its breath?

29th February

A day added to the year,
laconic or luminous.

The extra day can be seen
and touched, like any other.

Its hours are not difficult to count,
the weather varies but is weather,
no alien manifestation.

Lovers who marry on this day
have the usual eggshell hearts,
the lewdness of fish.

Children born on this day
are as fierce as any others.

Those who die on this day
must find new ways of being,

and on this day
singing still builds
the upstairs room of the sky.

This is the day
the year keeps for herself
but offers to you,
her breath for yours,
fair exchange.

The Divers

Yawning, the black hunters
check their guns,
fix their eye-masks.

They lumber over the ridged shale
with the composure of violinists,
virtuosi,

and stroll into the water,
the sea whispering like russian
in their chafed ears.

The two men relax in blizzards of fish,
hunt the small shark,

their harpoon guns banging lazily,
missing, spears chiming off rock;

the friendship of divers
has ghostly charm, gestures of sluggish
beauty; they drift like remote vacationers,

on a return trip to a voiceless woman
promising them a room of constant childhood,
its traditional clean sweep,

water mopping all corners with its forgetfulness,
swabbing away the beard-soft grime.

The men stride out of the sea,
indignant and separate.

The dress slipping out of his hands
holds its breath;
                    her room is changing the subject.

42

From Dreamland

(1, 3, and 7)

1   The Bed

Shall we lie down
like good children and sleep

in the undisturbed bed,
on quiet pillows,
on smooth unmarred white sheets?

All day the bright blue languages
of the sky
call out your name.

But we lie down for sleep
when the dark comes
with its separate fingers,
the nimble washer of our limbs.

3   Her Dream

According to an echo
it is the end of the year

but these stars
who are filling heaven

and shaking earth
with their legends,

who like to sleep in their clothes,

who have the bear's fierceness
and his lost lands,

who wear wishbone armour,
who play at marathons,

who can change colour like lizards,
whose dens and oases are in the dark,

these stars always keep something back,
something up their sleeves,
so that when you glance up again,

it is as if you asked your question
in the wrong language.

7 Dream Stairs

The long light boat skims the water,
stalking the current,

and the beasts on their pasture stairs
lift big hermit heads
to see who is coming.

The first stairs
are the blood-stairs,
blood never doubting its normal secrets.

Then, stairs carved out of rock,
quarry stairs
smelling of pepper and rain, berries and pearls.

There are bird-stairs translating messages
into ordinary language,
sensitive coherent workers;

there are stairs
to view the stars
half-hidden by weather and rosy moons;

there are needle-sharp stairs,
low-necked stairs;
steps of such beauty and dullness
that people refuse to climb them;

cliff-top stairs of the mainland,
polite stairs of the early morning,
steps that have stopped singing
or agreeing with gospels;

lightweight portable stairs,
kissable stairs,
whisper-stairs that promise, you will be rich . . .

Also,
stairs of cold and direct ways,
a black finger-arrow pointing, commanding,
driving you up the caustic stairs
of school, hospital, prison.

Stairs leading to interrogations.

Plus, precarious stairs of glass and water,
stairs you fall asleep on, stairs you fall down,
stairs that bring you to a passionate room
where a sea-damaged book lies open on a table
for you to read . . .

Dolls-house stairs too tiny to climb,
servant stairs of drudgery, your slow hands
drag the grey cloth over the wet stone.

Thick-carpeted regal stairs, for being queen,
or any other sacred person.

Wide shallow stairs for the child born blind.

Stairs that lessen the sufferings of the wounded.

The dream stairs are punctual,
made of polished wood shining like gold,
no speck of dust.

I climb all the dream stairs
to find out if my heart is mild or furious,
hollow or hot,
heart of a mother or a child.

The dream stairs with their death-traps,
their unexpected actors and immaculate music,
are more precise than birthday fictions.

You climb them to find your life, to claim it.

The swimmer lies flat in the tender water,
floating, her face submerged.
She is listening to water's dreams,
the nineteen dreams of water roll about her,
hammer gently on her willing head.

In the gossamer of the pool, my buoyant sister
swims another length, listening.

She has all the time in the world.

Hide and Seek

The child might be hiding in the ship
or in the cave,
or in the garden where the morning-glory
will find him some pretty name;
he might be hiding in the tree
whose shed needles fall like quills
on to the pitch of the dry lawn;
he might hide in a tower
built by a father for a son who never appeared,
the son dreamed-of but never caught up
into the real photographs of life;
the boy might hide by the cat-happy door,
or find some waterfall behind which to shelter,
be shuttered-in by the sheer fall of water;
he must hide somewhere.
He is a virtual prisoner in the powerhouse of the page,
must hide from the words thumping and beating on his head,
but where shall he hide, this boy who has not yet learnt
how to talk like a child,
or discovered that an evasive answer is the best way
to get uninterrupted possession of your day?
He hides everywhere, primitive, prodigal,
playing any number of odd games
in the garden of the red-eyed fish, their pool of stone and weed,
or in the stables where three horses watch him,
startled but, like electricity at rest, intensely patient.
The child hides, underfed in his blue shirt and french trousers,
in the room where I expected to find anyone else but him,
even a flock of those glossy and black gregarious birds
or the stately golden sane old dog of our crazy neighbours;
butter-finger room I at once let slip away into dullness,
losing him, he is not even behind those rivals, the curtains.

The child hid in a ship
and sailed away over an ocean, beneath deep-sea stars,
into the tenderness of storms,
the tempests, the burning calms,
the retentive and temperamental weather of a child
for whom no reward was offered and to whom nothing was promised.

The Lion from Rio

Golden inclination
of the huge maned head
as he rests against my knee,
his massiveness like feathers against me
amid this Rio crowd
through which he came to me,
this lion, my lion,
my lion of lifelong light,
padding unnoticed through the carnival.
Now his beast head rests in my lap,
golden flood, I am laden with it.
Looking up at me with his gentle puzzled gaze,
he says helplessly, but I am a man, a man!

My own child could have told me.
He was a man.
How could I not have seen it?
Listen again, he is drowsily moaning,
I am a man.

August Boy

The forest of summer is its own weight in gold
and you have climbed the tallest tree at noon
to bask above me and to kiss heaven, the fiery alpine.
High spirits! Silent golden child,
odd smiling pondering boy blossoming in a tree,
what country have you left to come and dwell here
in the burning branches and lion breath of the woods?
Child, boy, son I shall not carry, bear or nourish,
glowing ghost, summer boy who beckons to me
as I stand watching, soles of my feet scorching on the sandy path,
I know you are not a child I can claim,
you are not a child of the flesh, the fierceness of that.
Child without questions, child vigilant in a tree,
amazing as any thing made of gold, you live where the future is,
with all its carelessness and charm, its mistrust of direct answers.
The summer will not leave you behind, you are where summer is,
you are the heart of its heart, riding your solar beast,
the thoroughbred summer.
When I ask you your name you smile and say, 'you know my name'.
Furnace-Page of the Green and Gold of August,
Seigneur of the Summer, young Caesar of the Blazing Leaves,
wild, lenient and motherless, I recognize your boyish title.
Eagerly, easily, I lose my heart to you, my Heatwave Cupid.

The Eight Husbands

(traditional counting rhyme with cherry stones to foretell girl's future husband)

Our stove burns sea timber,
beach rags, a salty smoke.
Our stove burns newsprint brides,

bare bones of a dove,
long snakes of orange peel.
Our stove is never silent,
he sings and growls.

The child is sad, puzzled
by her ringed-finger,
its silver heart.

She weaves and fans her fingers,
listening to a tanglesome magic
I can't hear.

Is the ring a woman
and she still a child?
Is that the riddle?

But her sadness hurries away
when she rebukes it,
her heart does not dissolve,

not even experimentally,
its work is uninterrupted
by the childish unknowns
that visit her, outrage her.

On her winter nursery plate
there are black-red cherries
larger and sweeter
than any I have seen before.

The stove pops the cherry stones,
furnacing eight husbands in its kiln.

Sweet-mouthed, the child scrawls
written permissions on her page.

I heave the stove door shut,
wary of its blistering habits.

Somewhere salt waves are turning to stone.
A man is dancing on their strong floor.
Is he one of the eight?

Our room with walls of rose
and turmeric
smells of smoke and the turning tide,

smells of she, and I, and the little cat
who sleeps so courteously
on his green cushion of embroidered snowflakes,
their fleeting currency.

Down the mountain path
sneaks
the bigger cat.

Quiet bliss of his claws,
he moans
as he devours the young honeypot rabbit,
licks clean the shoelace bones.

Is this the finishing-stroke of evening?

We feed our guest the stove
again and again,
he is so hungry, this hound of fire,

he gobbles the garden twigs we gathered,
the wind-wrecked pampas grass,
snaps up our old exercise books and birthday cards,
how he loves his food,
puffs out noiseless smoke.

Above our roof it touches that hero, the sky.

Her husbands,
with their broad round demobbed heads,
their individual shoulders, legs
long or short, hands big or philosophical,
faces bare or bearded,
they are waiting for years to blow away
their smoke of months,
they wait higgly-piggly in the future, discussing tactics,
debts, addictions;

one in working boots, one in dancing shoes,
one in prison-dress lonely and fatigued,
one crowned with black helmet
and shadow visor concealing his face;
the fifth knows the horizons by heart,
the next has gypsy gold and gypsy horses
to captivate her,
another is sewing her twenty dresses to weep in,
and one has nothing at all to give her . . .

Who will she choose?

Their hearts roast in the stove.
Their names are sketched in the smoke.
They gleam, fiancé fish in grey water.
They practise their guessing games.

But she dreams
in her downcast chair of Egyptian blue
not of men

but of her white box of diamond dust,
her castles of wild rose and brocade,
her dancing rumours.

The child lifts her head,
looks at me across the room
through a fiery shock of sheaves,
girl's gold of single-seeded broom,
her long hair's suddenness.

What is her question to be?

My guesswork darkens the room.

Sleepily I refill her dish of cherries.
Silently she accepts,
kicks off her shoes,
sits without payment in the shadows.

We eat the fruit,
we tend the fire,
we watch the animals,
we count the men.

She and I can be fitted together
in a multitude of ways to make our tangram,
our mother/daughter pattern.

We live in this room, cradled,
soothed by it, rocked
as if we lived in Canton
on a boat house.

The child will allow me
to forget nothing
that happens to us here.

Her husband will remember also.
That is his flattery, his future,
his rainy-day wedding task,
recalling this room, its life as shepherd,
as star, as rose.

Life Line

(*palmist*)

1

Blue sky falls into your open hand,
life's razor busy on your palm.

Blue wounds, fate's groove,
love's gashes, the grazes of fortune.

The left hand is happy,
the right hand waits.

This laborious lane of grief
and brutality is the head line.

Love's line is a whiplash of hearts,
taut magic, accidental learning
by touch or dream or fragrance.

(A hand not open to the public.)

Blue sky painting its labyrinth on your hand,
blue sky making your hand his mirror,
looking for his fortune.

2

The hand has to come home alone,
frailed with its lines.

Soft mitten of skin
written on indelibly.

Back of the hand
lonely as silver.

The heart line trembles above the life line,
like a cloud over a mountain.

The life line knows
the heart has to come home alone.

The life line's route is long and hellish.
But to it the heart makes its salaam.

Acknowledgements

Some of these poems first appeared in the following magazines and anthologies: *Poetry Australia*, *The Manhattan Review*, *The Scotsman*, *The Literary Review*, *Poetry Chicago*, *Argo*, *Ambit*, *The New Statesman*, *Verse*, *Outposts*, *Words*, *PEN New Poetry 1*, and *No Holds Barred*.

My thanks are due to the Arts Council of Great Britain for financial assistance during the preparation of these poems.